Classifying Living Things

Printed in Mexico

ISBN-13: 978-0-15-362073-7

ISBN-10: 0-15-362073-0

2 3 4 5 6 7 8 9 10 805 16 15 14 13 12 11 10 09 08

Harcourt
SCHOOL PUBLISHERS

Visit *The Learning Site!*
www.harcourtschool.com

VOCABULARY

classification

How Are Organisms Classified?

Grouping living things with other living things like them is called **classification**. Scientists use characteristics to classify organisms. They group the snake with other animals as part of the animal kingdom. They classify the orchid as part of the plant kingdom.

Why Living Things Are Classified

Sorting things into groups of similar items is called **classification.** When you go into a library, you see that all similar books are in one area. The librarians classify the books so you can find the book you want. For example, all the fiction is in one part of the library. This is an example of classification. Just as libraries classify books, scientists classify living things.

This coral reef in Florida is home to many different kinds of living things.

Scientists have classified about 2 million different kinds of living things, or organisms. Classifying organisms makes it easier for scientists to study them.

Classification helps scientists in two ways. First, it helps them correctly identify organisms. Second, classification helps scientists better understand how living things are related to one another.

Living things can be classified in many ways. Scientists have to agree on one system so that when they discuss an organism by name, they all know what organism it is.

Aristotle had a classification system. He classified all living things as either a plant or an animal. Then he classified animals by the way they moved. The dragonfly belongs with the fliers, the centipede belongs with the walkers, and the octopus belongs with the swimmers. How would humans be classified in Aristotle's system?

As scientists find new information, classification systems change. Scientists can now use high-powered microscopes to study organisms. They can also study the DNA of organisms to classify them. DNA study can help scientists identify differences between organisms. This helps them classify the organisms.

 Why do scientists need one system of classifying living things?

Scientists compare DNA patterns to classify organisms and identify relationships.

The system we use today is based on Carl von Linné's system of organization.

7

How Living Things Are Classified

Carl von Linné's system of classification is called the Linnaean system. Von Linné based his system on what organisms looked like. He divided living things into two groups—the animal kingdom and the plant kingdom.

How would von Linné classify the green tree snake? The orchid?

Today, scientists use other characteristics to classify organisms. For example, they can use cell structure and ways of getting food.

Scientists have also discovered living things that are different from most plants and animals. These organisms do not fit into the plant kingdom or the animal kingdom. So scientists have added other kingdoms.

One common classification system has six kingdoms. As new discoveries are made, classification systems will continue to change.

 What characteristics do scientists use to classify organisms?

Euglena have chloroplasts like a plant. But they move like an animal. Euglena are classified in a kingdom that has been added to the Linnaean system. ▶

Review

Complete this main idea statement.

1. A _____ system of living things grows and changes as scientists make new discoveries.

Complete these detail statements.

2. _____ ____ _____ developed the classification system we use today.

3. Scientists use _____ _____ and ways of getting food to classify organisms.

4. There are six _____ in today's classification system.

What Are the Major Groups of Organisms?

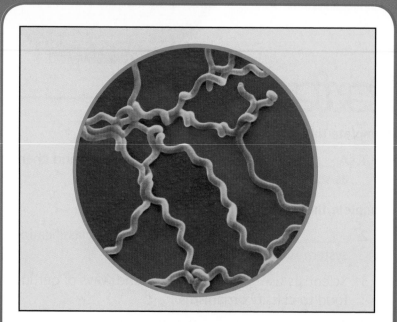

Bacteria are single-celled organisms that lack nuclei.

A **fungus** is an organism whose cells have walls but no chloroplasts.

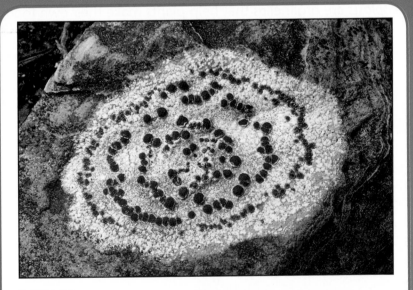

Protists are very tiny organisms that may have characteristics of plants, animals, or fungi. Slime molds, above, have some of the characteristics of fungi.

COMPARE AND CONTRAST

You **compare and contrast** when you look for ways things are similar and different. **Compare** means to find the way things are similar. **Contrast** is to find the way things are different.

Compare and contrast how the major groups of organisms are alike and different.

Plants

Have you ever walked in a park during the summer? What color do you mostly see? You probably see a lot of green. You walk on green grass, touch green leaves on trees, and see green shrubs and mosses. Green is in all of the living plants in the park.

The plant kingdom includes many different organisms. It includes organisms as small as mosses and as big as trees. Remember that plants have cells with walls. Plant cells also have chloroplasts. *Chloroplasts* are organelles that help plants make their own food.

▼ The pine tree belongs to a group of plants called *conifers.* Conifers make new trees from seeds formed in cones.

Scientists can classify plants by the way they reproduce, or make more of themselves. They can also classify them by their structures—roots, stems, leaves, and flowers.

Scientists also look at the adaptations of plants. An adaptation is a feature of an organism that helps it live in its surroundings. Some plants have an adaptation called *vascular tissue*. Vascular tissue is tubelike tissue. Water and nutrients move through the plant in vascular tissues. Scientists classify plants into two groups by whether or not they have vascular tissue.

 How is a vascular plant different from a nonvascular plant?

This flowering plant has vascular tissue. It is a vascular plant. It reproduces from seeds formed by flowers.

This moss does not have vascular tissue. It is a nonvascular plant. Mosses do not make seeds. They reproduce by forming spores. New plants grow from the spores.

Animals

Animal cells are different from plant cells in two important ways. Animal cells do not have cell walls, and they do not have chloroplasts.

Scientists classify animals by whether or not they have a backbone. Animals with a backbone are called *vertebrates*. Vertebrates are classified into smaller groups because of their body coverings, the way they breathe, and how they reproduce. The five groups of vertebrates are fish, amphibians, reptiles, birds, and mammals.

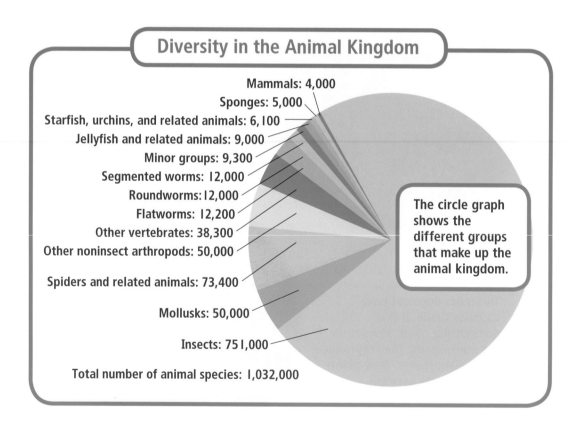

Diversity in the Animal Kingdom

Mammals: 4,000
Sponges: 5,000
Starfish, urchins, and related animals: 6,100
Jellyfish and related animals: 9,000
Minor groups: 9,300
Segmented worms: 12,000
Roundworms: 12,000
Flatworms: 12,200
Other vertebrates: 38,300
Other noninsect arthropods: 50,000
Spiders and related animals: 73,400
Mollusks: 50,000
Insects: 751,000
Total number of animal species: 1,032,000

The circle graph shows the different groups that make up the animal kingdom.

◀ Animals with a backbone are called *vertebrates.*

Animals without a backbone are *invertebrates.* Invertebrates make up more than 95 percent of all animals. Scientists classify invertebrates with skeletons into one group. They then classify this group into smaller groups based on several questions: Does it have a body cavity? What is its body shape? How does it digest food? Does its body have sections that repeat?

What characteristics do scientists use to classify vertebrates into smaller groups?

Animals without a backbone are *invertebrates.* ▶

Fungi

A **fungus** is an organism whose cells have cell walls but no chloroplasts. At one time, scientists classified fungi as plants. But as they learned more about fungi, they classified fungi in their own kingdom. Most fungi eat the decaying tissue of other organisms. Some fungi can cause disease in other living things.

Scientists classify fungi by their shape, size, and way of reproducing. They also classify them by how many cells they have. Yeasts are single-celled. Most other fungi have many cells. Mushrooms are fungi with many cells.

Scientists also classify fungi by where they live, how they live, their cell structure, and their reproductive systems.

 How are fungi similar to plants? How are fungi different from plants?

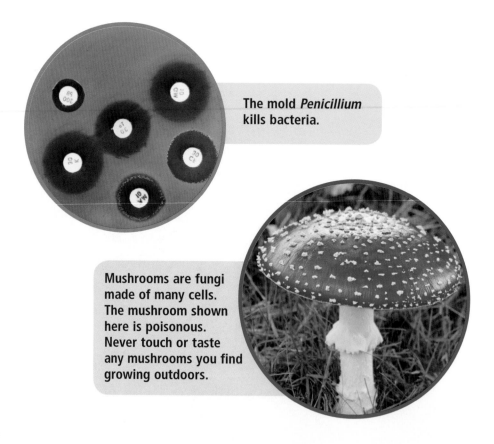

The mold *Penicillium* kills bacteria.

Mushrooms are fungi made of many cells. The mushroom shown here is poisonous. Never touch or taste any mushrooms you find growing outdoors.

Protists

Protists are very tiny organisms that may look like plants, animals, or fungi. Most protists have only one cell. Each cell has a nucleus with DNA.

Scientists classify protists by whether they are more like animals, plants, or fungi.

Plantlike protists have chloroplasts. They can make their own food. Plantlike protists are mainly grouped by color. They are also grouped by the types of materials they use to store food.

Scientists classify animal-like protists by their size, shape, and the way they move. An *amoeba* is an animal-like protist. It does not have a cell wall or a definite shape. It moves by using "false feet." The false feet form when cytoplasm flows against the cell membrane. A *paramecium* is another animal-like protist. It moves by using hairlike parts called *cilia*. The cilia let the paramecium "swim."

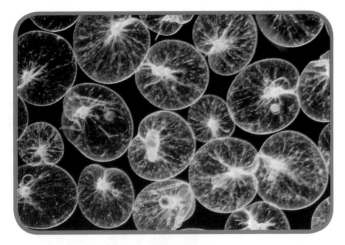

▲ These organisms, commonly called sea sparkles, leave a glowing trail in seawater. Like plants, they make their own food.

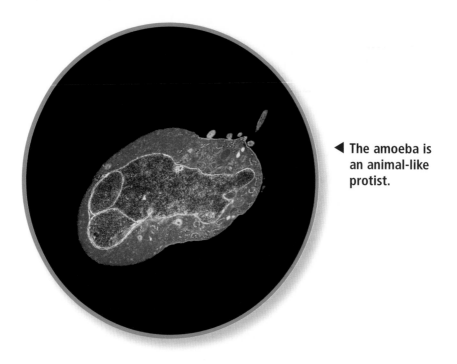

◀ The amoeba is an animal-like protist.

One kind of funguslike protists are called *slime molds.* The cells have cell walls. But they do not have chloroplasts. Slime molds are grouped by their shape and how they reproduce.

 How are plantlike protists similar to plants?

Funguslike protists are also called *slime molds.* ▶

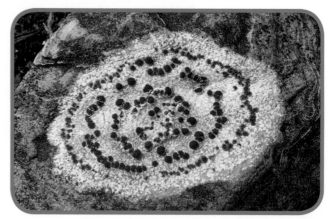

Bacteria

Bacteria cells do not have a nucleus. This makes them different from the members of all the other kingdoms.

Scientists divide bacteria into two kingdoms. One kingdom includes bacteria that live in very harsh environments, such as hot springs or near deep-ocean vents. The other kingdom includes all other bacteria.

Bacteria live almost everywhere on Earth. Most kinds of bacteria are harmless, or even helpful. Bacteria live in your body. You need them for body processes. Very few bacteria cause disease.

Scientists classify bacteria by size and shape, how they get energy, and whether or not they can make food.

 What are the two kingdoms of bacteria?

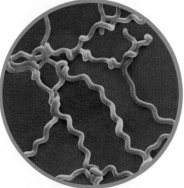

These bacteria with a twisted shape are classified as spirochetes. ▶

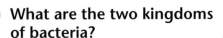

Focus Skill

Complete these compare and contrast statements.

1. Plants, such as flowering plants, with tubelike structures are called _____ plants.

2. A vertebrate is an animal _____ a backbone. An invertebrate is an animal _____ a backbone.

3. Bacteria are not like other organisms because their cells do not have a _____.

4. Organisms whose cells have walls like plant cells but no chloroplasts are called _____.

How Do Scientists Name Organisms?

A **genus** is a group of organisms that share major characteristics. They are closely related. The genus *Felis,* for example, includes the ocelot and the house cat.

A **species** is a group of organisms that can reproduce live offspring of the same kind. These zebras are members of one species. The wildebeests are a different species.

MAIN IDEA AND DETAILS

The **main idea** is what the text is mostly about. **Details** are pieces of information about the **main idea**.

Look for information about how scientists name organisms and **details** about the naming process.

Grouping and Naming Organisms

Suppose you need a book for a school report. You know how to find it in your school library, but you have to go to your local public library. You would be able to find the book here too. This is because most libraries use the same system of classification.

Scientists use a classification system too. They use their system to help them talk about organisms. They used to have problems talking about organisms. This was because the same organism might have different names in different languages. It made it difficult for scientists to work together. Von Linné made a classification system to solve this problem.

▼ The scientific name for the giraffe is *Giraffa camelopardalis.*

Number of Species in Some Phyla of the Animal Kingdom		
Number	**Phylum**	**Includes**
1,000,000	Arthropoda	insects, spiders, crabs
50,000	Mollusca	oysters, snails, squid
43,000	Chordata	fish, mammals, birds
9,000	Annelida	earthworms, leeches
6,000	Echinodermata	starfish, sea urchins

Von Linné grouped organisms into several levels. He gave each level a Latin name. All scientists understood Latin.

The largest level is the *kingdom*. The two smallest levels are *genus* and *species*. A **genus** is a group of organisms that share major characteristics. They are closely related. A **species** is a group of organisms that can reproduce live young of the same kind.

An organism's classification determines its scientific name. A scientific name has two parts. The first part tells the genus. It begins with a capital letter. It is usually written in italics. The second part tells the species. A species name begins with a lowercase letter. It is also written in italics.

For example, cats are part of the genus *Felis*. *Felis* is Latin for cat. A house cat is called *Felis domesticus*. A mountain lion is *Felis concolor*. They share the same genus name, so you know house cats and mountain lions are closely related.

Scientists use the scientific names of an organism to be sure that they are all talking about the same animal. The names also may show how organisms are related to one another.

 Why did scientists need one system of scientific naming?

Using the Linnaean System

In the Linnaean system, living things are classified into seven levels. At each level, all of the organisms share certain characteristics. Look at the chart below. It shows the classification of the common house cat. Its scientific name is *Felis domesticus*.

The higher levels have more animals. These animals all have some things in common. As you move down the chart, you can see that there are fewer animals on each level. But animals on the lower levels have more in common with one another. At the lowest level, the animals have the most in common.

The lowest level is species. Members of a species all share at least one characteristic that no other animals have, and they can mate with each other.

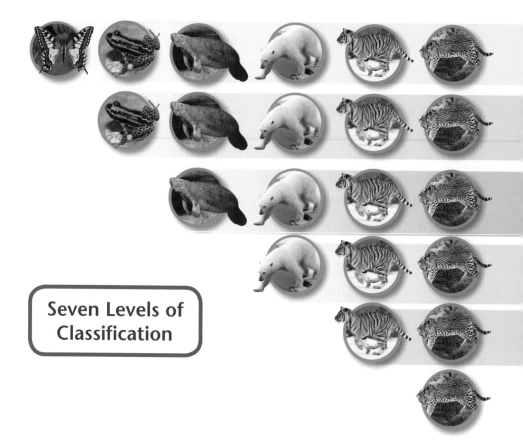

Seven Levels of Classification

An organism's classification tells you a lot about the organism. The group it belongs to at each level tells you some of the organism's characteristics.

For example, the house cat belongs to the class *Mammalia.* All animals in this class have fur, bear live young, and feed their young with milk. So all of these characteristics are true of the house cat.

The cat also belongs to the order *Carnivora.* All of these animals eat meat. So the cat eats meat.

Because the cat belongs to the family *Felidae,* you know that it has some characteristics in common with both the ocelot and the tiger.

 Do all animals in the family *Felidae* eat meat? How can you tell?

Kingdom: The kingdom Animalia, or animals, is divided into phyla. The phyla include chordates, mollusks, worms, jellyfish, sponges, and arthropods, such as *Danaus plexippus,* the Monarch butterfly.

Phylum: The phylum Chordata is made up of all vertebrates and some animals that have a cartilage rod instead of a backbone. It is divided into classes of mammals, fishes, birds, reptiles, and amphibians, such as *Dendrobates auratus,* the poison dart frog.

Class: The class Mammalia is made up of animals that have fur or hair, bear live young, and produce milk for their young. It is divided into orders, including carnivores, primates, elephants, rodents, bats, and sirenians, such as *Trichechus manatus,* the manatee.

Order: The order Carnivora is made up of meat-eating animals. It is divided into families, including cats, dogs, weasels, seals, and bears, such as *Ursus maritimus,* the polar bear.

Family: The family Felidae, the cat family, consists of the genus *Felis* and the genus *Panthera*—cats that roar, such as *Panthera tigris,* the tiger.

Genus: *Felis domesticus* belongs to the genus *Felis,* purring cats, which also includes such cats as *Felis pardalis,* the ocelot.

Species: House cats belong to the species *Felis domesticus.*

Using a Key

Have you ever looked at an animal and wondered what it was? Even scientists who study nature do not know the name of every animal. They use guides to help them identify the animal.

Some people who go hiking or camping take guides that help them identify animals. One type of guide is called a field guide. A field guide is a book with pictures or drawings of organisms. It gives the name of the organism next to the picture or drawing. Some field guides also provide a brief description of the organism, including where it is generally found.

Another kind of guide is called a *dichotomous key*. Dichotomous means "divided into two parts." The key is made up of a number of steps. Each step has two options. Only one of the options describes the organism you want to identify. You decide which one. Your choice takes you to another pair of options. You keep going until you get to the name of the organism.

 How could you use a field guide to help you identify an animal?

Heron Identification Key

1a. Long yellow legs . . . . Go to **2**
1b. Short yellow legs . . . . Green heron,
Butorides virescens

2a. Long, pointed bill . . . . Go to **3**
2b. Long thick bill . . . . Great blue heron,
Ardea herodias

3a. White chest and rust neck . . Tricolor heron,
Egretta tricolor

Review

Focus Skill

Complete this main idea statement.

1. Living things are organized using the _____ system so that scientists around the world can work together.

Complete these detail statements.

2. A group of organisms that share major characteristics and are closely related is called a _____.

3. A group of organisms that can reproduce among themselves is called a _____.

4. A _____ _____ is made up of choices that guide a user to the name of an organism.

GLOSSARY

bacteria (bak•TEER•ee•uh) Single-celled organisms that lack nuclei.

classification (klas•uh•fih•KAY•shuhn) The sorting of things into groups of similar items.

fungus (FUNG•guhs) An organism that has cell walls but does not have chloroplasts.

genus (JEE•nuhs) A group of organisms that share major characteristics and are, therefore, closely related.

protist (PROH•tist) A very tiny organism that may have characteristics of plants, animals, or fungi.

species (SPEE•sheez) A single group of organisms that can reproduce among themselves.